LOOKING AT CHROMOSOMES

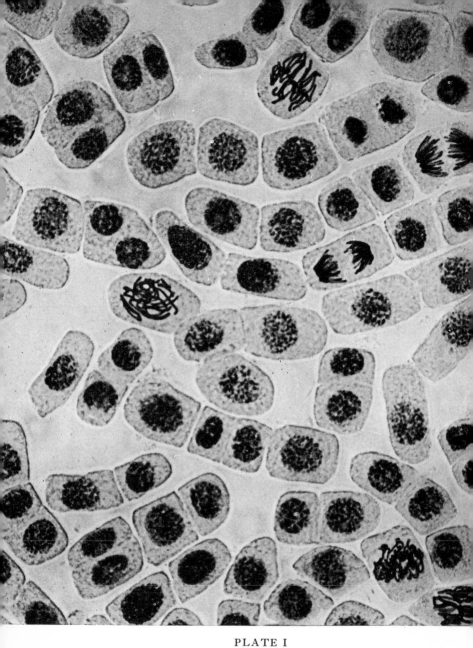

PLATE I

Cells from part of a root-tip, prepared by the Feulgen squash method,
seen at a low magnification ($\times 480$).

LOOKING AT CHROMOSOMES

by

JOHN McLEISH, Ph.D.

and

BRIAN SNOAD, M.I.Biol.

John Innes Institute, Colney Lane, Norwich

SECOND EDITION

MACMILLAN
ST MARTIN'S PRESS

First published 1958
Reprinted 1959, 1962
Reprinted with minor amendments 1966
Second edition 1972

Published by
THE MACMILLAN PRESS LTD
London and Basingstoke
Associated companies in New York Toronto Melbourne
Dublin Johannesburg and Madras

Library of Congress catalog card no. 70–85144

SBN 333 13790 6

Printed in Great Britain by
ROBERT MACLEHOSE AND CO. LTD
The University Press, Glasgow

PREFACE TO THE FIRST EDITION

P LANTS and animals are made up of cells which contain the substances concerned in heredity. We can begin to understand the mechanism whereby these substances are transmitted from one generation to another from a detailed study of cells under an ordinary high-power microscope. Our aim in this book has been to demonstrate by means of a continuous series of photographs and illustrations, how cells and the hereditary substances they contain behave during the growth of a living organism, how the cells which function in reproduction come to be formed and, finally, how the hereditary substances are passed on from parents to offspring.

There are few living organisms in which all the various phases of cell division are suitably clear for the purposes of demonstration. It has, therefore, been usual to combine observations from a range of plants and animals, each one favourable for the demonstration of particular stages. In our opinion, however, cell behaviour is more easily understood if the observations are confined to one species so that a less interrupted sequence can be obtained. There are some species of flowering plants which are suitable in this respect and it is from one of them, *Lilium regale*, that we have been able to obtain our series of photographs. We are, therefore, primarily concerned with demonstrating the behaviour of the chromosomes in plant cells although we must emphasise that the underlying mechanism of heredity is fundamentally similar in plants and animals.

The book is not intended to be a comprehensive survey of cell behaviour but we hope that the combination of photographs, drawings and text will prove of value to those students with some knowledge of biology who wish to know something of chromosomes. Those who have already started a course of practical cytology may find the photographs of some use in the interpretation of their own preparations.

We have found the Feulgen squash method the most suitable staining technique for demonstrating the chromosomes. Techniques for demonstrating other cell components such as nucleoli, spindles and cytoplasmic bodies are available but if they had been used many of the details of chromosome behaviour would have been obscured.

As no elaborate equipment is needed preparations such as ours can easily be made and we urge those who are interested to make some for themselves. A description of cytological and photographic technique is not within the scope of this book but details can be found in one of the books which we recommend (see Appendix).

We should like to thank Mr. L. S. Clarke, Photographer at this Institute, not only for his helpful advice during the taking of the photographs, but also for his careful preparation of the prints.

J. McL.
B. S.

BAYFORDBURY
January, 1957

PREFACE TO THE SECOND EDITION

THE series of photographs is still the main feature of the book and remains unaltered but the text has been expanded, where necessary, to incorporate additional information. More details are now given of the chemical nature of the cell's genetic components, and differentiation and cellular variability are more fully discussed. Some facts are also given about the timing of the mitotic and meiotic cycles in relation to plant development. The Appendix now includes details of the techniques employed in preparing the microscope slides and photographs, and the list of books for reference and further reading is extended.

I have, unfortunately, to record that the death of my colleague Dr. John McLeish occurred during the final stages of the preparation of this new edition.

NORWICH   B. S.
July 1971

CONTENTS

1

INTRODUCTION

UNTIL the end of the sixteenth century theories concerned with heredity, reproduction and growth were speculative and frequently based upon superstitions and traditional beliefs. As microscopes with higher magnification and improved resolution gradually became available, however, it became possible to examine the component parts of living organisms in far more detail. Step by step new discoveries were made and these slowly dispelled many of the superstitious ideas which had arisen in the past.

By the end of the seventeenth century it was realised that both plants and animals varied in appearance because the living material of which they were composed could be organised in a variety of ways. Hooke, Grew, Malpighi and others showed that this living material was not a single homogeneous mass but was split up into numerous microscopical units called **cells,** although it was well over one hundred years before it was discovered that growth was dependent upon the behaviour of cells. A single cell was seen to give rise to two new cells apparently by dividing in half and soon it was realised that growth depended upon such **cell division;** cells could multiply and the pattern in which they subsequently became arranged determined the appearance of the mature organism. It was generally agreed that cells could arise only by the divisions of pre-existing cells and, moreover, that every plant and animal started its life as a single cell. It was, however, some years before the full implications of these facts were appreciated.

With the aid of more powerful microscopes greater attention was paid to the contents of cells and in 1831 Brown discovered that every cell contained a spherical body which he called the **nucleus.** Eleven years later Nägeli noticed that at certain stages in the life of a cell the nucleus became transformed into a number of smaller bodies which were later given the name **chromosomes** by Waldeyer. It was Waldeyer who pointed out that the transformation of nucleus into chromosomes must be highly significant as it always accompanied the reproduction or division of a cell.

Towards the end of the nineteenth century biologists began to realise that growth of the tissues of an individual was correlated with cell behaviour and that the division of a cell was always accompanied by regular changes of the nucleus. The cell and its contents were seen to be involved, not only in the growth of an individual, but also in the process of sexual reproduction. In 1875 Hertwig showed that in animals the fusion of the nuclei of two special cells — the **sperm** from the male and the **egg** from the female — gave rise to a new individual. This was the process of **fertilisation.** Two years later Strasburger recorded fertilisation in plants: a process basically similar to that observed by Hertwig in animals except that the cells involved in plants were the pollen grains and embryo sacs.

At last it was realised that the nucleus was the carrier of the hereditary material and that a new individual arising from sexual reproduction must contain hereditary material from both parents. Gradually the information, which had been collected through the years by various independent observers, led to a clearer conception of heredity. Weismann was one of the first to correlate the behaviour of cells and their contents with what was known of heredity at that time and he also postulated a mechanism whereby the inherited material could be passed on from one generation to another by means of the chromosomes.

While these important advances in **cytology** were being made numerous other experiments were being conducted into the inheritance of characters in plants and animals. Those experiments carried out by Mendel proved to be the most important because they demonstrated the precise way in which some characters were passed from parents to offspring. In addition they showed that with a knowledge of the characters peculiar to an individual it was possible to predict with a great degree of accuracy how these characters would be inherited. Mendel's contemporaries did not fully appreciate the importance of his conclusions which were, therefore, soon forgotten. In 1900, however, three independent workers, De Vries, Correns, and Tschermak, published the results of their experiments which were basically similar to those published by Mendel thirty-five years earlier. Thus, **Mendel's Laws of Heredity** became widely accepted and formed the foundation of the new science of **genetics**. Within three years it had been shown by Sutton that there was a striking parallel between the Mendelian Laws of Heredity and the behaviour of chromosomes in the formation of germ cells. At last the true relationship between heredity, growth and reproduction was established. The functional units of heredity, called **genes,** were found to be arranged in a linear order along the thread-like chromosomes and although they were too small to be seen the presence of each gene could be confirmed by its specific activity.

It was not until the early 1940's that Beadle and Tatum, using a fungus called *Neurospora*, demonstrated that genes were responsible for the production of proteins, some of which were the enzymes involved in the regulation of cell activity. This was the first of many remarkable discoveries related to the chemistry of cells and chromosomes, but before describing what is undoubtedly the most important discovery, brief mention must be made of work carried out as far back as 1869. Just three years after Mendel's results were published Miescher demonstrated that nuclei contained a chemical compound which he termed nuclein. It is now known that nuclein contains an important molecule, **deoxyribonucleic acid** (**DNA**) which today is recognised as the substance of the gene. The significance of Miescher's findings, like those of Mendel, was not appreciated at the time and more than seventy years passed before it became evident that DNA might well be the fundamental genetic component. DNA was found in the nuclei of all living organisms and, moreover, it was found in constant amounts in nuclei of an individual. Most significantly perhaps, Avery, MacLeod and McCarty showed in 1944 how DNA extracted from one bacterial strain could induce permanent heritable transformations in another strain.

Watson, Crick and Wilkins soon provided a vital piece of information which had hitherto been lacking. From the results of chemical and physical analysis they were able to construct a model which demonstrated how the components of DNA — the purine and pyrimidine bases, sugars and phosphates — could be arranged in an orderly way which gave the molecule the form of a long, double helix. Perhaps more significantly, they suggested a mechanism whereby the molecule could undergo self-replication. The model was widely accepted and provided final confirmation for a link between the chemistry of DNA and the behaviour of chromosomes.

Needless to say, this work has given tremendous impetus to biological research in general and, in particular, to research concerned with the way in which genetic information encoded in DNA is transcribed and translated during various forms of cell activity.

A full grasp of these modern concepts of gene structure and activity obviously requires some knowledge of physics, chemistry and mathematics. But a basic understanding of the way in which the genetic material is transmitted during cell division and reproduction is essential and this, at least, can be obtained from looking at chromosomes in plants such as *Lilium regale*.

MITOSIS AND DEVELOPMENT

FLOWERING plants which reproduce by sexual means develop from a single cell, called a **zygote,** which is the product of fertilisation. The development of this cell into a mature individual starts with its division, or to be more precise its reproduction, into two new cells. These two cells may then divide to give four and, similarly, the four can divide to give eight (Fig. 1). The process is repeated many times although the divisions may not continue to occur synchronously and only a few cells may be dividing at any one time.

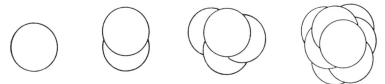

FIG. 1. Cell multiplication. Showing, diagrammatically, how eight cells may be derived from one cell. In this way a zygote develops into a young embryo.

During the course of development of the plant from the embryo through the seedling stage to maturity, the cells continue to multiply until there are many millions of them. In *Lilium regale,* for example, it can be calculated that a mature plant is made up of about one hundred million cells and this is a relatively small number for a species of higher plant. Because all these cells arose from one cell, the zygote, they obviously have something in common. In fact, the parts of the cell which are concerned in heredity, the chromosomes and genes, are present in every cell in exactly the same form and amount. Each cell, therefore, contains the same molecules of DNA and has the same hereditary potentialities. This is a most significant fact and one which is important,

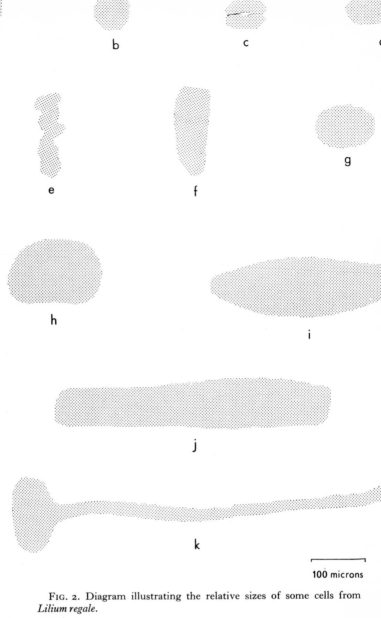

FIG. 2. Diagram illustrating the relative sizes of some cells from *Lilium regale*.

(a) Root meristem cell.
(b) Pollen mother cell.
(c) Pair of stomatal guard cells.
(d) Leaf epidermal cell.
(e) Leaf palisade cell.
(f) Cell from stem cortex.

(g) Pollen grain.
(h) Cell from bulb scale parenchyma.
(i) Embryo-sac mother cell.
(j) Cell from differentiated region of root.
(k) Root hair.

100 microns

not only for the normal life of the plant, but also for that of its descendants in subsequent generations.

In the early stages of development from the zygote, the embryonic cells are relatively uniform in size. They soon come to differ from one another in size and shape, however, as they begin to fulfil different functions in the various tissues which are forming. Ultimately, when the tissues in a mature plant are examined under the microscope, cells are seen which differ enormously not only in shape and structure but also in mass and volume (Fig. 2). Cells serve a variety of specific functions. Some may help to confer mechanical strength on the plant while others may be involved in protection, the uptake of nutrients, photosynthesis, storage, transport or reproduction. It seems paradoxical, therefore, that such diversity can result from a uniformity of nuclear composition; but it is on the maintenance of this continuity that the species depends for its normal development. Such continuity is ensured by a process called **mitosis** (Fig. 4) in which the division and reproduction of a cell take place in a very precise and orderly manner.

In the growing points of stems and roots there are tissues called **meristems** which are centres of active cell multiplication (Plate I). In these it is possible to examine mitoses in great detail and because of this we are now familiar with the main visible changes.

The cells of meristematic tissues which are not actually in the process of division are said to be in a stage called **interphase**. The interphase cell can be regarded as a fundamental working unit with three principal component parts as illustrated in Fig. 3. Each has its own wall which, in plants, is a fairly rigid structure composed of a microfibrillar framework of cellulose embedded in a gel-like matrix. If such cells are grown in isolation, as can be done in carefully maintained cultural conditions, they assume a roughly spherical shape. When forming an integral part of a tissue, however, they are typically polyhedral since they are in close contact with several other cells.

Inside the cell wall is the **cytoplasm** which can be seen under the electron microscope to be enclosed by a membrane. It is difficult to define the structure of the cytoplasm when using a light microscope but it has now been revealed in some detail by electron microscopy. It consists of a system of folded membranes, known as the endoplasmic reticulum, associated with thread-like

B

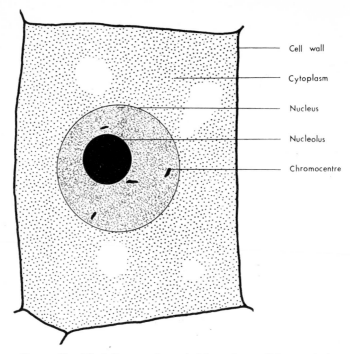

Cell wall

Cytoplasm

Nucleus

Nucleolus

Chromocentre

FIG. 3. Simplified diagram of a typical interphase cell from actively
growing plant tissue.

and microtubular structures. Also present are mitochondria,
Golgi bodies, ribosomes, lysosomes and plastids each concerned
with particular types of synthesis. The nucleus is perhaps the
most conspicuous component of the cell and it is enclosed by a
porous double membrane, visible under an electron microscope,
and the outer part of this membrane is often seen to be contiguous
with other cytoplasmic membranes particularly those of the endoplasmic reticulum.

While it must be borne in mind that it is the chemical interaction of nucleus and cytoplasm which determines the pattern of
behaviour of the interphase cell, it is important to focus attention
on the nucleus for it is here that the visible changes associated
with mitotic division will occur.

During interphase the nucleus is composed of a number of
greatly attenuated thread-like **chromosomes** most parts of which
cannot be distinguished under the light microscope. The chromo-

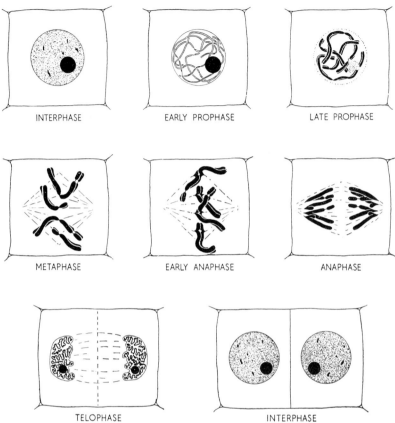

FIG. 4. Illustrating the main stages of cell division by mitosis in plants.

somes at this stage can be resolved with the electron microscope, however, and are often seen to be dispersed in a finely granular matrix or nuclear sap. In some plants, small dense-staining regions of some of the chromosomes do, however, remain clearly visible under the light microscope during interphase: these are known as **chromocentres**. In addition, every nucleus contains one or more bodies called **nucleoli** which are formed close to small chromosomal regions termed **nucleolar organisers**. In any given species, the number of nucleolar organisers per nucleus is constant but the number of nucleoli may be reduced, at any time during interphase, because of their tendency to coalesce. The results both of specific staining reactions carried out on intact cells and of chemical analysis of isolated nucleoli have shown that these organelles have many components of which proteins and nucleic acids are the most prominent. Although nucleoli are dense spherical bodies with well-defined boundaries, no outer membrane is present and electron microscopy has shown them to consist of a compact mass of fine threads and granules.

Nucleoli are believed to play an important part in nucleic acid metabolism and protein synthesis and they may also be the centres for regulating the passage of genetic information from the DNA of the chromosomes. It is a significant fact that they are always very prominent in actively growing meristematic and embryonic cells as well as in secretory cells in which large quantities of protein are being made.

Interphase cells are engaged in many important synthetic activities which influence and are influenced by the activities of similar cells in the surrounding tissue. The most important of these, from this present point of view, are those activities which culminate in mitosis and enable the cell to produce two copies of itself. Although it is not fully understood what stimulates an interphase cell to undergo mitosis, the process must be preceded by the synthesis of new protein and the replication of DNA, the two main components of the chromosome. It is difficult to give the precise length of time occupied by interphase in plants as it varies from cell to cell and also depends upon environmental conditions, particularly temperature. Usually, under optimal conditions, it takes about 20 hours (Fig. 5) of which approximately the middle one-third (S) is spent in DNA synthesis and in the synthesis of a basic protein, **histone,** which is closely associated with the DNA. Chromosomal histone, like DNA, is precisely doubled in amount during the S period. There are two gaps in interphase, immediately before and after the S period, and these are known as G_1 and G_2 respectively. Other nuclear proteins and a substance called **ribose nucleic acid (RNA)** which is allied to DNA, are synthesised throughout G_1, S and G_2. This synthesis certainly contributes to the general increase in cell and nuclear mass which usually precedes mitosis.

Mitosis begins with a stage called **prophase** (Plates III and IV and Fig. 4). The chromosomes which constitute the nucleus become clearly visible under the light microscope as long threads

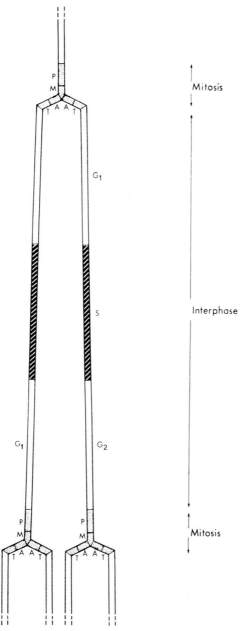

FIG. 5. Diagram illustrating the relative duration of the phases of the mitotic cycle. In *Lilium* one complete cycle usually occupies some 18 to 24 hours.

P, prophase; M, metaphase; A, anaphase; T, telophase; G₁ pre-synthesis gap; S, period of DNA synthesis; G₂ post-synthesis gap.

each of which is divided along its length into two identical halves called **chromatids**. The chromatids of each chromosome are often twisted around one another. The chromosomes are single threads at the beginning of interphase yet when they become distinguishable at prophase, each appears to have reproduced; this reproduction occurs during the S period of interphase when the chemical components of the chromosome double in amount. Each chromosome thus replicates and becomes two chromatids identical in size, shape and chemical composition. There is one small region of each chromosome, called the **centromere,** which is seen to be undivided at this stage. It is not until much later that each centromere is seen to divide and play its highly speci-alised role.

As prophase continues the chromosomes gradually become more distinct due to a complicated system of coiling, and possibly folding, which causes them to contract and thicken (Plates V and VI). While this contraction continues, the nucleoli gradually dis-appear. The fate of their components is unknown although there is evidence to suggest that some may be transferred to the chromo-somes. The rest are probably dispersed in the nuclear sap and cytoplasm.

About the time that the chromosomes reach maximum con-traction, the nuclear membrane breaks down and the contents of the nucleus mingle with the cytoplasm. This is the end of prophase.

The next stage, **metaphase,** is heralded by several changes in both the cytoplasm and the chromosomes. Once the nuclear mem-brane has broken down a new structure begins to develop consist-ing of a large number of delicate fibres orientated roughly parallel to one another; because of its general shape this structure is known as the **spindle.** The extreme ends of the spindle are called the **poles** and the central region, the **equator.**

From the results of chemical analysis and observations made with the polarising and electron microscopes, it is now known that the spindle fibres are microtubules of approximately 15–20 nano-metres outer diameter. Cross-sections of individual microtubules show that they are commonly composed of thirteen protein sub-units arranged in a circle.

The chromosomes now move very gradually through the cyto-plasm and become arranged on the equator in a special way. The centromere region of each chromosome, which is still the only undivided portion and is marked by a constriction, comes into contact with the spindle. The centromeres are always arranged in one plane across the equator but the free arms of the chromosomes are not necessarily restricted to any one position. By this time the chromosomes have attained their maximum degree of contraction. Metaphase has now been reached (Plate VII and Fig. 4).

Metaphase is the most favourable stage for the examination of individual chromosomes. As has already been seen each cell is a derivative of the zygote and the zygote is in turn derived from the fusion of nuclei from the parental germ cells. It is not surprising to find, therefore, that at the metaphase of mitosis there are two sets of chromosomes: one from each of the parents. In other words each chromosome has a *morphologically identical* partner. The number of chromosomes in each of the parental sets is known as the **haploid** number, while the two sets together in the zygote, and hence in all the cells of the plant, constitute the **diploid** number. It should however be realised that, owing to their different origins, the chromosomes of each pair *may not be genetically identical* although, on morphological grounds, they are usually said to be homologous pairs. The morphologically identical partners may have alternative forms of some of the genes: **alleles.** Such differences in the DNA code are, of course, not visible under any microscope.

The detailed examination of chromosomes at metaphase is facilitated by the use of certain drugs, the most effective of these being colchicine which is an alkaloid obtained from *Colchicum autumnale*, the Autumn Crocus. Such drugs inhibit the formation of the spindle so that the chromosomes lie scattered freely in the cytoplasm instead of forming a compact group in one region of the cell. The chromosomes become excessively contracted and the various constrictions are more easily seen because there is less chance of overlapping (Plate VIII).

Every chromosome has a centromere, sometimes known as the **primary constriction,** which occurs at a position that is constant for and characteristic of that particular chromosome. Some chromosomes have, in addition, a **secondary constriction** (Fig. 6). Some secondary constrictions mark the positions of the nucleolar organisers which were mentioned in the description of the interphase nucleus. The constancy in position of both primary and

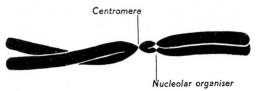

Centromere

Nucleolar organiser

FIG. 6. One of the A chromosomes of *Lilium regale* showing the two types of constriction (*cf*. Plate VIII and Fig. 7).

secondary constrictions, together with the variation in size and appearance of the chromosomes of the complement, provides a useful key for the identification of species and particularly of hybrids (Fig. 7).

The next stage in the mitotic cycle is **anaphase**. The centromeres, which at metaphase were the only undivided parts of the chromosomes, now divide so that each chromatid has its own centromere. This change marks the very beginning of anaphase. The division of the centromeres appears to take place at the same time in all the chromosomes and the sister centromeres thus formed then appear to repel one another so that they move apart. This movement results in a drawing apart of the sister chromatids

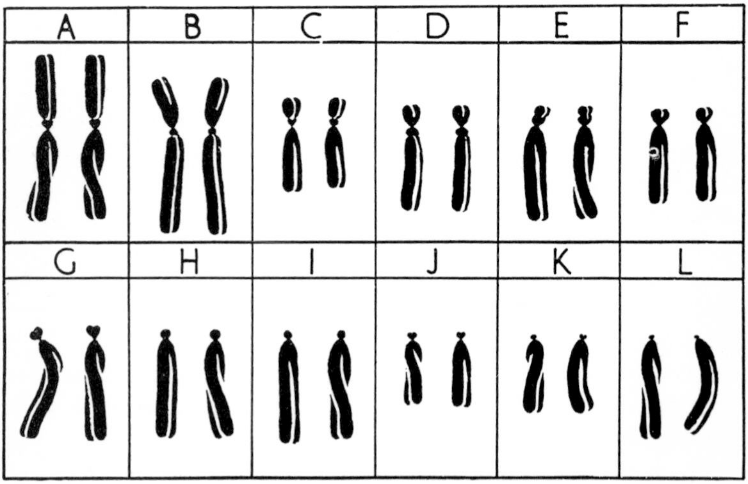

A	B	C	D	E	F
G	H	I	J	K	L

FIG. 7. Drawings of the 24 metaphase chromosomes of *Lilium regale* arranged as 12 homologous pairs with their centromeres, or primary constrictions, in line. Secondary constrictions present in A, B, C, D and E (*cf*. Plate VIII).

in a progressive separation which starts at the centromere region and continues along the chromatid arms towards the ends (Plates IX and X and Fig. 8). Improved methods in time-lapse cinematography and in the techniques for keeping cells alive in artificial culture media, have enabled observation of the speed of anaphase movement and the mode of chromatid separation. The movements seem to be governed by the changing shape of the spindle and by the contraction of its fibres or microtubules which result in the drawing apart of the sister chromatids. Commonly, in plants, this takes place at a rate of about one micrometre per minute.

FIG. 8. Centromere division leading to the separation of the sister chromatids.

The sister chromatids, or chromosomes as they can now be called, continue their slow movement towards opposite poles of the spindle (Plates XI, XII and XIII and Fig. 8). At this later stage of anaphase the spindle lengthens and becomes narrower at the equator; such lengthening of the spindle is said to complete the anaphase separation of the daughter chromosomes which now form two closely packed groups, one at each spindle pole. This marks the end of anaphase and the beginning of the next stage, **telophase** (Plates XIV and XV).

During telophase the chromosomes behave in a manner which is, in many respects, the reverse of that seen during prophase. The chromosomes uncoil to become long thin threads and nucleoli begin to form. The chromocentres, so characteristic of many species of plants and animals, soon become evident, while the chromosomes begin to form a typical interphase nucleus with the development of a nuclear membrane. These changes occur at exactly the same time in both daughter nuclei which are still within a common cytoplasm. Now, however, at the equator of the spindle a new cell wall begins to form which separates the two nuclei and roughly divides the cytoplasm of the original cell in two.

Mitosis is now complete and two new cells have been formed (Plate XVI). Looking back over the successive phases through which the chromosomes have passed it is now possible to appreciate the purpose and significance of the mechanism of mitosis.

During mitosis every chromosome reproduces itself longitudinally and this reproduction of the chromosomes is, of course, accompanied by the replication of the genetic component, DNA. When the identical sister chromatids separate at anaphase, this genetic component is transported to opposite ends of the cell and it is here that the two daughter nuclei are formed. Such precise division can be confirmed by the use of special microphotometric techniques in which the actual amounts of DNA are measured in individual nuclei.

Thus the nucleus of each daughter cell comes to contain exactly the same genes as the nucleus of the original cell from which it was derived and in this way genetic continuity is ensured.

Each new cell may follow one of two main patterns of development. On the one hand, it may enter another interphase, replicate its DNA and undergo a further mitosis. On the other hand, it may not divide again before becoming modified in such a way as to fulfil any one of a number of specialised functions in the organism. Such specialisation, or **differentiation,** may itself be accompanied by one or more further periods of DNA synthesis to give in each nucleus, for example, two, four, eight or sixteen times the original DNA content without accompanying cell division. It is difficult to understand how various tissues and organs come to diverge so widely in both form and function when all the cells of a mature organism are derived from the zygote and are thus potentially genetically identical. In plants, the formation of leaves, stems, roots, flowers and other organs is brought about by differentiation. It is clear that the properties of development and differentiation are inherited because they are expressed in the same way in each successive generation. It is also significant that isolated parts of plants in the form of root, stem and leaf 'cuttings', as well as undifferentiated cells and pollen grains on synthetic media, are capable of regenerating a new plant with characters identical to those of the plant from which they were originally taken.

The results of a variety of biochemical and genetical experiments are beginning to demonstrate how the nucleus controls organised growth through two remarkable properties of its DNA. DNA makes copies of itself by replication but, in addition, it can act as a template in making some forms of RNA which pass from the nucleus to the cytoplasm. These RNA molecules, carrying copies of the genetic 'message', then help to regulate metabolic activity by directing the syntheses of specific proteins at appropriate sites in the cytoplasm. The ability of cells to metabolise, to undergo mitosis and to become integrated with other cells in the formation of a particular tissue, is the result of control by specific regions of the DNA molecules which are activated or inactivated at appropriate times during the life of the organism. In any one cell, at any given point in time, only a small proportion of the genetic DNA is likely to be active, the rest being 'masked' by proteins, particularly the histones. At the same time, different regions may be 'unmasked' and active in other cells which are performing other functions. The way in which this differential activation is brought about is still unknown but there is no doubt that it depends on the exchange of substances between neighbouring cells as well as on the chemical interaction between nucleus and cytoplasm.

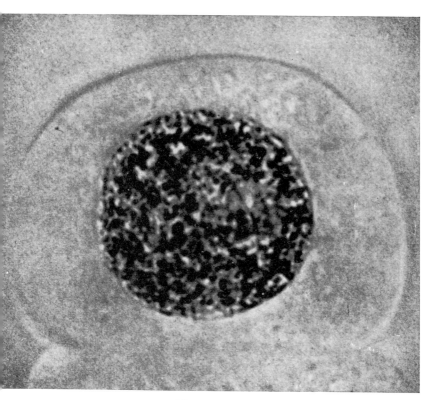

PLATE II

Interphase in a root-tip cell. The chromosomes are not individually distinguishable (× 2400).

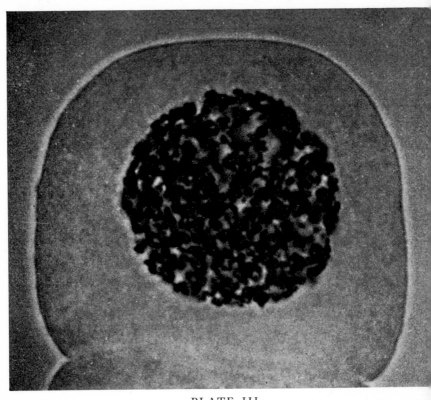

PLATE III
Prophase. The chromosome threads are just visible (× 2400).

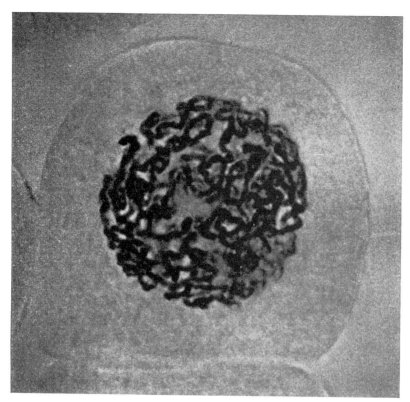

PLATE IV

Prophase. The chromosome threads are now quite distinct (× 2400).

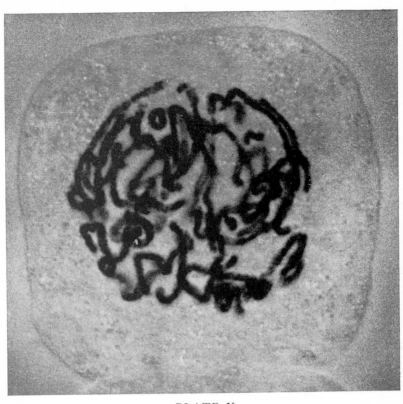

PLATE V

Prophase. The chromosome threads are shorter and thicker. Each is
composed of two chromatids (× 2400).

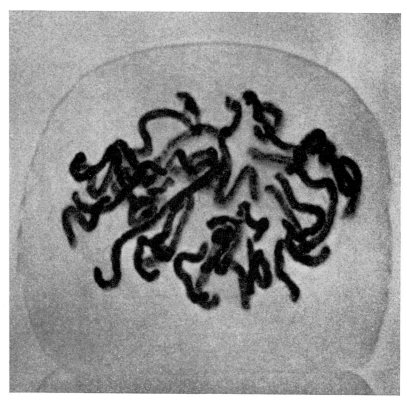

PLATE VI

Prophase. The chromosomes are reaching their maximum degree of contraction and the nuclear membrane is breaking down (× 2400).

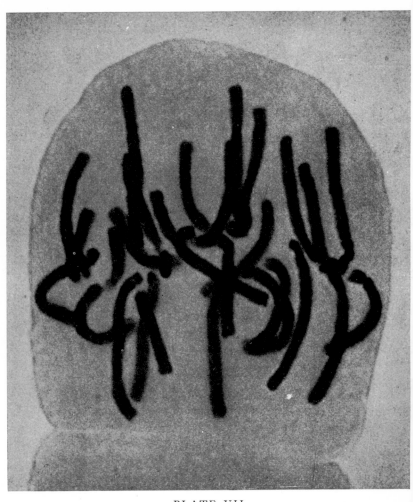

PLATE VII

Metaphase. The chromosomes have contracted further and they are now arranged upon the equator of the spindle. The spindle is not visible when the Feulgen staining method is used (× 2400).

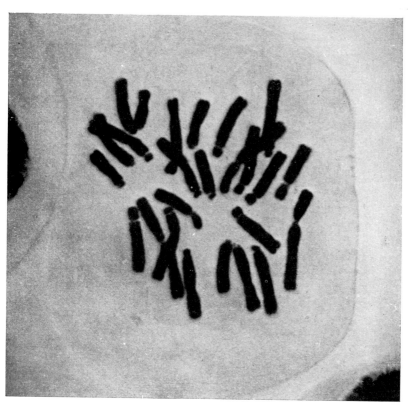

PLATE VIII

Metaphase. After colchicine treatment the 24 chromosomes and their centromeres, or primary constrictions, are clearly seen (× 2400).

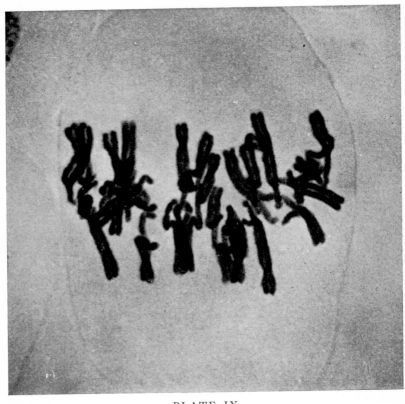

PLATE IX

Anaphase. The chromosomes are still arranged on the spindle equator but their centromeres have divided and are beginning to move apart (× 2400).

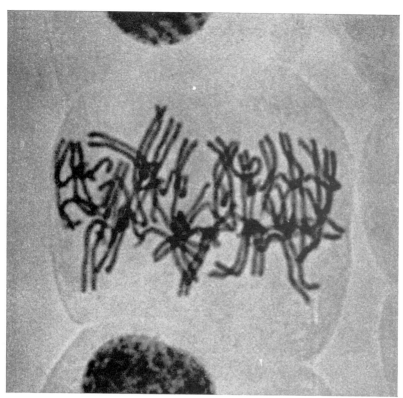

PLATE X

Anaphase. The centromere movement is continuing and the chromatids
are being separated (× 2400).

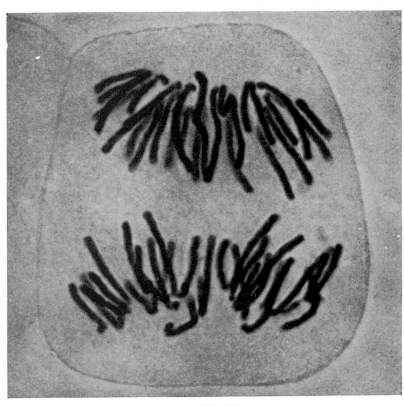

PLATE XI

Anaphase. The daughter chromosomes have separated and they are
moving to opposite poles of the spindle (× 2400).

PLATE XII
Anaphase. The centromeres have reached the poles (× 2400).

PLATE XIII

Anaphase. The daughter chromosomes are forming two distinct groups, one at each pole of the spindle (× 2400).

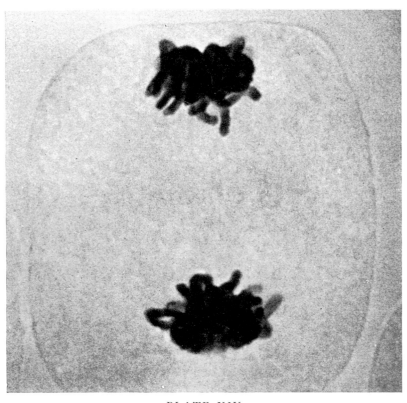

PLATE XIV

Telophase. The chromosomes are beginning to form the two daughter nuclei (× 2400).

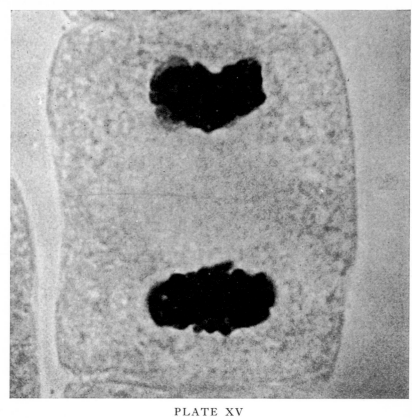

PLATE XV

Telophase. The chromosomes within the two daughter nuclei are no longer distinguishable. A cell wall is beginning to form between these two nuclei (× 2400).

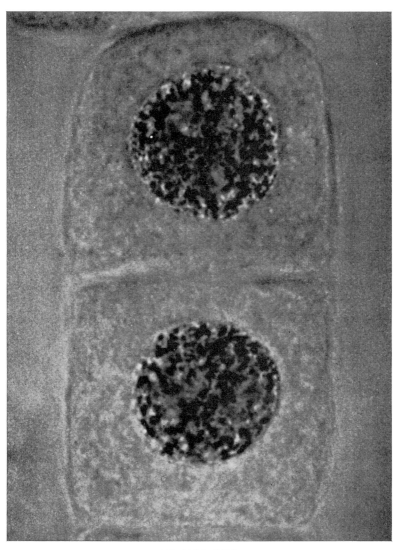

PLATE XVI

Interphase. Two new nuclei have now been formed which are completely separated by the new cell wall (× 2400).

3

MEIOSIS AND REPRODUCTION

As a plant develops towards maturity certain tissues become differentiated for the purpose of reproduction. This may be of the vegetative or asexual type where bulbs, corms, tubers and similar organs are produced. But in the majority of flowering plants sexual reproduction leading to the formation of the seed is of more common occurrence.

Sexual reproduction involves the formation of specialised male and female germ cells in the development of the **pollen grains** and **embryo-sacs** respectively. It is the fusion of special kinds of nuclei from these male and female cells, in a process known as **fertilisation,** which gives rise to a zygote, the first cell of the new individual. If the male and female nuclei each contained the same number of chromosomes as the other cells in the plant, that is the diploid number, an impossible situation would arise through fertilisation because in each successive generation the chromosome number would be doubled. A special process has evolved, presumably by the gradual modification of mitosis, whereby this difficulty is overcome. This process is called **meiosis** and it comprises two divisions resulting in the formation of germ cells whose nuclei contain the haploid chromosome number (Fig. 9). This is half the number found in the diploid cells of the rest of the plant.

The synthesis of DNA, RNA and proteins has been studied in meiosis as well as in mitosis. It has been shown that the amount of DNA, the material of the genes, is the same in cells just prior to the meiotic division as it is in cells which are preparing to undergo mitosis. DNA is synthesised in each nucleus before the beginning of meiotic prophase and it is eventually distributed to the four haploid germ cells.

(i) *The Formation of Pollen Grains*

The pollen is formed in the anthers of the young developing flower buds. In early stages of development the anthers contain a

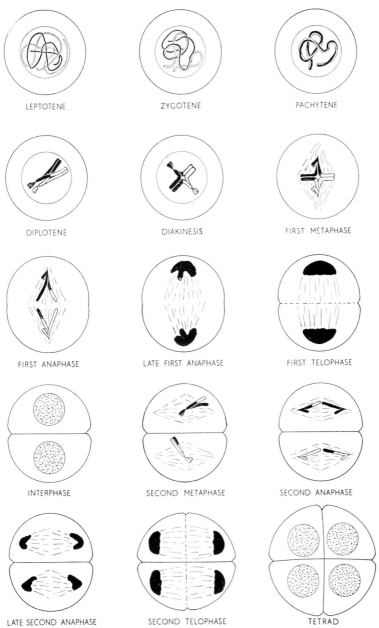

LEPTOTENE

ZYGOTENE

PACHYTENE

DIPLOTENE

DIAKINESIS

FIRST METAPHASE

FIRST ANAPHASE

LATE FIRST ANAPHASE

FIRST TELOPHASE

INTERPHASE

SECOND METAPHASE

SECOND ANAPHASE

LATE SECOND ANAPHASE

SECOND TELOPHASE

TETRAD

Fig. 9. Diagram illustrating the main stages of meiosis in plants. Only one maternal chromosome (white) and one paternal chromosome (black) represented, for the sake of clarity.

mass of diploid cells undergoing numerous unsynchronised mitoses. These are at some later stage subject to a stimulus whose nature is unknown but which results in greater synchronisation (Plate XVII). The degree of synchronisation in these cells varies considerably between different species of plants but when it has reached a certain stage and the **pollen mother cells** are formed meiosis begins. This involves two nuclear divisions but only one division of the chromosomes in each pollen mother cell. Consequently each of the four resulting nuclei contains one half of the number of chromosomes present in the original nucleus.

Prophase is the first stage of meiosis. It is a longer and more elaborate stage than the prophase of mitosis and is sub-divided into the following stages: **leptotene, zygotene, pachytene, diplotene** and **diakinesis.**

Within the pollen mother cells at **leptotene** the nucleus is composed of delicate, thread-like chromosomes which, because of their extreme length, are not individually distinguishable under the light microscope (Plate XVIII). At first sight leptotene resembles the prophase of mitosis but closer examination can reveal important differences. First, the chromosomes are *single* and not divided longitudinally into two chromatids as at mitosis; and, secondly, their structure is more definite. They somewhat resemble strings of beads, because of dense granules, called **chromomeres,** which occur at irregular intervals along their length. Chromomeres have characteristic sizes and positions on each chromosome and because of this they act as crude markers for the approximate positions of genes or groups of genes.

During leptotene and throughout the entire prophase of meiosis the chromosomes remain enclosed in a well-defined nuclear membrane. In appropriately stained preparations the nucleoli are very distinct during this stage and are much larger than they are during the prophase of mitosis. These nucleoli are always, as in mitosis, associated with the specific regions of particular chromosomes called the nucleolar organisers and it is assumed that the nucleolar material is synthesised as the result of their genetically controlled activity. The chromosomes are present in the diploid number as they are in most other cells of the plant. In other words, there is one set of chromosomes which originally came from the male parent and an homologous set from the female parent.

As the next stage, **zygotene,** is reached these homologous chromosomes begin to move. Each chromosome in one parental set has a partner of similar size and form in the other parental set and these pairs of chromosomes begin to come together. Pairing of the homologous chromosomes begins at one or more points and gradually proceeds along their whole length in a 'zipper-like' fashion (Plate XIX and Fig. 10). It should be emphasised that

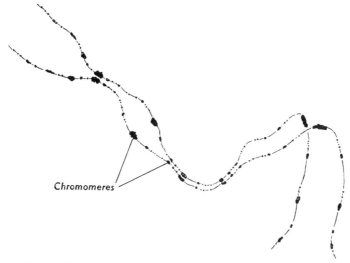

Chromomeres

FIG. 10. Part of a pair of homologous chromosomes at zygotene. Pairing between homologous regions has started (*cf*. Fig. 9).

pairing is a very exact process with chromomeres of similar size and at corresponding positions on each chromosome being brought together. While this is taking place the chromosomes are also undergoing the shortening and thickening which is typical of prophase whether it be in mitosis or meiosis. Pairing is so complete that the nucleus eventually appears to contain the haploid number of chromosome threads. Only the most critical observation under the high-power light microscope shows each thread to consist of a pair of homologous chromosomes called a **bivalent**. Further evidence for the pairing of chromosomes at this stage comes from electron micrographs which show what appear to be chromosome regions in close association. These associated regions are called **synaptinemal complexes.**

During the next stage of prophase, **pachytene,** the bivalents

D

become even shorter and thicker and sometimes individual bivalents or parts of bivalents can be recognised by the characteristic arrangement of their chromomeres (Fig. 11). The centromere
region and the region with which the nucleolus is associated can
sometimes be seen and often act as further useful guides to the
identification of bivalents.

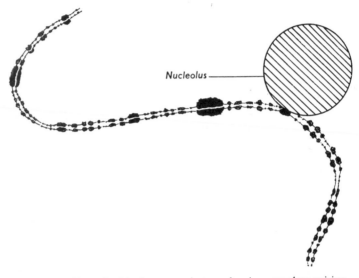

Nucleolus

FIG. 11. Part of a bivalent at pachytene showing complete pairing
between the homologous chromosomes (*cf*. Fig. 9).

The paired chromosomes of each bivalent become more intimately associated and, as pachytene proceeds, they often appear to
become twisted around one another. As is to be expected the bivalents themselves will not associate with one another (Plate XX).

There are other changes at this stage which are not visible but
which have far-reaching genetical consequences and it can be in-
ferred that these changes have taken place in pachytene from the
appearance of the chromosomes at the next stage of prophase
which is known as **diplotene**.

A striking change in the appearance of the bivalents marks the
beginning of diplotene. The attraction between the homologous
chromosomes in each bivalent now begins to lapse so that they are
no longer in close contact along their entire length. There are,

however, one or more places in each bivalent, called **chiasmata,** where contact is retained (Plate XXI). In particularly favourable cells it can now be seen that each chromosome has divided along its entire length except at one region, the centromere. The bivalent now consists of two pairs of chromatids. The chromosomes of each bivalent remain held together at the chiasmata because of a union which has formed at homologous regions. This is a special kind of union between chromatids, not a sticking together but an exchange of genetic material, at precisely corresponding positions.

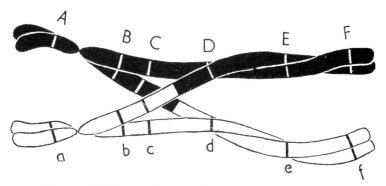

FIG. 12. A bivalent at diplotene with a single chiasma. Some of the genes are represented by letters. Crossing-over has occurred between one chromatid of the maternal chromosome (white) and one chromatid of the paternal chromosome (black).

In contrast to zygotene and pachytene, where there appeared to be a force of mutual attraction between homologous chromosomes of each bivalent, there now appears to be repulsion which results in the formation of loops between the consecutive chiasmata along the bivalent. Further contraction of the chromosomes takes place during this stage; the chromomeres become less distinct and the chromosomes themselves assume a rather 'woolly' appearance. But, before proceeding further with the description of meiosis, it is necessary to consider the consequences of **chiasma formation.**

Each chiasma provides visible evidence that there has been an exchange of parts of chromatids between the paternal and maternal chromosomes of each bivalent. The exchange itself is often referred to as **crossing-over** and the reason for this should become clear from Figure 12. If it is assumed that the sequence of genes in part of *one chromatid* involved in the chiasma can be represented by

A B C D E F

and in the corresponding part of the *other chromatid* by

a b c d e f

Then, if a chiasma is formed which involves the regions between C and D and c and d, the genes in each chromatid will now become arranged as follows:

A B C d e f

and

a b c D E F

The actual mechanism whereby this exchange of chromatid parts, or crossing-over, is brought about must depend upon the reproduction of each of the parental chromosomes forming the bivalent and the time at which it takes place. Unfortunately, the time of reproduction cannot, as yet, be fixed with any degree of certainty although it probably occurs before the actual double nature of the chromosomes is apparent under the microscope. Until more is known, therefore, of the composition and behaviour of nucleic acids and chromosomes during the early stages of meiosis, crossing-over is likely to remain a subject of great controversy. There is, however, no dispute about the fact that genes are recombined in this way by the process of crossing over.

Although the paired chromosomes in each bivalent are, in most cases, morphologically homologous it must be remembered that if they originated from different parents they could well carry different complements of genes and the extent of this difference will depend upon the ancestry of each individual. Plants that reproduce by self-pollination are more likely to be genetically uniform and it is only by out-breeding, that is by crossing two genetically dissimilar plants, that there will be genetic differences between the chromosomes of each pair. Crossing of genetically dissimilar plants and the consequent re-assortment of genes at meiosis is of prime importance in evolution.

Each parental chromosome carries its share of the total complement of genes and these genes are arranged, as we have already stated, in a definite linear sequence upon each chromosome. This arrangement is, normally, strictly maintained by mitosis during the development of the individual, and it is only during meiosis that there can be a reciprocal exchange of chromatid parts so that new combinations of genes can arise. The extent of such gene recombination will depend upon the number of chiasmata in each bivalent and this in turn is dependent upon the length of the chromosome. The distance between successive chiasmata is quite large in relation to the length of the chromosome and thus the number of chiasmata occurring in each bivalent is comparatively limited. All the genes on a chromosome form what is called a **linkage group**. Comparatively simple mathematical analysis of the frequencies of gene recombination make it possible to determine not only how individual genes are arranged within the linkage groups but also how far apart they are from each other. This type of information is of great value to the geneticist undertaking practical breeding work.

Before the genetical implications of meiosis can be discussed any further, it is essential to see how the chromosomes, which have been modified by crossing-over, behave in the final stages of meiosis and how they come to be distributed among the pollen grains.

Diplotene passes gradually into the next stage, called **diakinesis,** during which the paired chromosomes of the bivalents reach maximum contraction (Plate XXII). As contraction proceeds it can be seen that the consecutive loops between chiasmata come to lie in planes at right angles to one another. This gives the appearance of links in a chain.

The positions now occupied by the chiasmata do not necessarily represent the original points at which crossing-over occurred. The coiling and contraction of the chromosomes together with the repulsion between them leads to movement of the chiasmata towards the ends. This movement or slipping of the chiasmata is known as **terminalisation** (Fig. 13). By the end of diakinesis, which is also the end of meiotic prophase the nucleolus and the nuclear membrane are no longer visible. A spindle is formed and the bivalents begin to move slowly through the cytoplasm towards the equator. Each bivalent consists of two chromosomes and thus each bivalent possesses two centromeres. These centromeres do *not* come to lie in one plane parallel to the equator of the spindle, but, instead, the bivalents become orientated on the spindle so that their two centromeres lie one on each side of the equator and at equal distances from it. This stage is known as the **first**

DIPLOTENE

DIAKINESIS

METAPHASE

FIG. 13. Showing the progressive changes in a bivalent, with two chiasmata, from diplotene to the first metaphase of meiosis. Contraction and thickening of the chromosomes accompany terminalisation of the chiasmata. At diakinesis and metaphase, therefore, the chiasmata no longer mark the positions at which crossing-over originally occurred.

metaphase of meiosis (Plate XXIII and Fig. 9). The orientation does not necessarily result in all the centromeres of the maternal chromosome set lying on one side of the equator and all the centromeres of the paternal chromosome set lying on the other. The arrangement is completely random.

The orientation of the centromeres together with the additional contraction and coiling of the chromosomes results in a striking difference between the metaphase of mitosis and that of meiosis (Plates VII and XXIII).

If one bivalent is particularly easy to distinguish from the others because of a striking difference in size, its appearance can be studied in different pollen mother cells. It can then be seen that the number and position of chiasmata vary from cell to cell and modify the shape of the bivalent accordingly. Some variation of chiasma frequency and position between pollen mother cells is quite characteristic of all species.

The next stage is the first **anaphase of meiosis**. The two centromeres of each bivalent which remain undivided begin to move away from each other towards the opposite poles of the spindle and at the same time the chromatids are gradually loosened and the chiasmata start slipping further along towards the ends of their chromosomes (Plate XXIV).

As the centromeres continue to move towards the poles they pull their half-bivalents with them (Plates XXV and XXVI). The rate at which the half-bivalents separate depends upon the length of the chromosome arms and the number of chiasmata which have to be drawn apart. The shape of the spindle begins to alter, as it does in a mitotic anaphase, and this accompanies the final separation of the half-bivalents to the poles. It must be remembered that these half-bivalents are chromosomes in which at least one chromatid contains some recombined sequence of genes. This is the **first telophase of meiosis** and now the chromosomes become uncoiled and form two nuclei which are usually separated by a cell wall which forms in the region of the spindle equator (Plates XXVIII and XXIX). The two cells are together known as a **dyad**.

It should be clear that after the first division of meiosis the chromosome number in each of the two nuclei is half that of the original nucleus of the pollen mother cell. Each now contains the haploid number. Moreover, these nuclei could differ genetically for two reasons. First, as a result of crossing-over the combination of genes on each chromosome may have been altered. Second, even if there were no crossing-over the random orientation of the maternal and paternal centromeres could, at anaphase, lead to genetical differences between the two daughter nuclei.

The two nuclei then pass into what is called **interphase** (Plate XXX and Fig. 9). This varies considerably in duration in different species of plants. In some cases it is merely a transitory phase with the chromosomes remaining clearly visible throughout. In other cases the chromosomes may form a nucleus which resembles that seen at mitosis. It should be pointed out here, however, that there are important differences. In mitosis the daughter chromosomes which go to form the interphase nucleus are single threads. On the other hand, the half-bivalents which go to form the interphase nucleus at the end of the first division of meiosis are clearly double threads, having divided prior to diplotene, and they are still held together by their undivided centromeres. There is no further reproduction of the chromosomes during the interphase of meiosis nor is there any synthesis of DNA.

This interphase is a prelude to the last part of meiosis known as the second division. This division serves to separate the genetically different chromatids of each chromosome. At the end of the first division a cell wall is usually laid down between the two nuclei of the dyad. It is unlikely that the two nuclei will enter **second division** simultaneously when they are separated by such a cell wall. Where there has been no distinct interphase and the individual chromosomes have remained visible, the first telophase of meiosis passes imperceptibly into the **prophase** of the second division. In those organisms where the chromosomes uncoil at first telophase and become indistinguishable within the meiotic interphase nucleus the prophase of second division starts with the re-appearance of these chromosomes (Plate XXXI). In either case, as prophase proceeds, the chromosomes contract, thicken, become more distinct and can then be seen to be pairs of threads which are held together by a still undivided centromere. They have, in other words, remained structurally unaltered since the end of the first division.

At the end of prophase a spindle appears in each cell of the dyad. The orientation of the two spindles relative to one another varies within and between different groups of plants (Fig. 14).

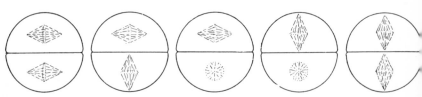

FIG. 14. Dyads at the second anaphase of meiosis showing some of the possible ways in which the two spindles can be orientated relative to one another.

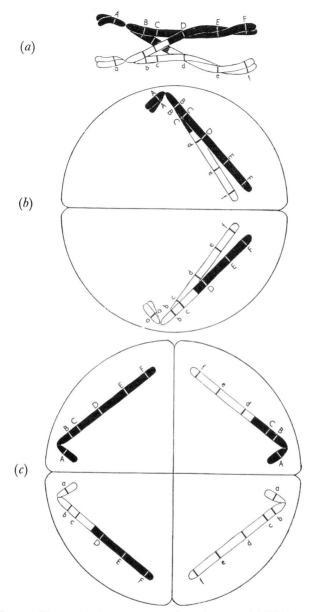

(a)

(b)

(c)

FIG. 15. The genetical consequences of crossing-over: (a) A bivalent at diplotene with a single chiasma. Some of the genes are represented by letters.
Crossing-over has occurred between one chromatid of the maternal chromosome
(white) and one chromatid of the paternal chromosome (black); (b) The half-
bivalents in the dyad after the first division of meiosis; (c) The chromosomes in
the four potential germ cells, each differing from the others genetically, after the
second division of meiosis.

As in mitosis the chromosomes become arranged on the equator of the spindle (Plate XXXII). This is **second metaphase**. The centromeres divide at the beginning of **second anaphase** and the daughter chromosomes move to the poles (Plates XXXIII and XXXIV). These daughter chromosomes are equivalent to the chromatids in the original pollen mother cell but are possibly recombined as a consequence of crossing-over during chiasma formation. At **second telophase** there are four nuclei which eventually become separated from one another by new cell walls (Plates XXXV and XXXVI). Each cell is destined to become a pollen grain and the four cells are collectively known as a **tetrad** (Plates XXXVII and XXXVIII). Four potential pollen grains are formed each with a haploid set of chromosomes and each could differ from the others genetically, as shown in Figure 15, if the diploid pollen mother cell was heterozygous.

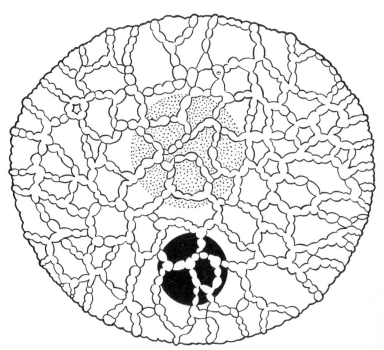

FIG. 16. Pollen grain of *Lilium regale* showing the characteristically patterned wall. The first pollen grain mitosis has resulted in the formation of a vegetative nucleus (stippled) and a generative nucleus (black).

Eventually, the four pollen grains separate, grow and begin to develop a thick wall which is often characteristically patterned (Plates XXXIX and XL). After a period of continuous growth the nucleus of each pollen grain undergoes a mitotic division but the time at which this **first pollen mitosis** takes place varies for different pollen grains within an anther because the nuclei of the pollen grains are genetically different. As a result two nuclei are formed, the **vegetative nucleus** and the **generative nucleus**. Although these nuclei are genetically identical they soon differ widely in appearance, the generative nucleus becoming compact while the vegetative nucleus becomes diffuse and indistinct (Fig. 16). This is regarded as an example of differentiation *within* the cell.

While these changes are occurring within the pollen grains, the growth of the flower bud continues until eventually it opens to expose the stamens and carpels. Throughout meiosis and during the early stages of pollen development the contents of the anthers are somewhat fluid. Now, however, rapid drying occurs and eventually the anthers split longitudinally to expose the pollen grains which may then be dispersed by movements of the air or by

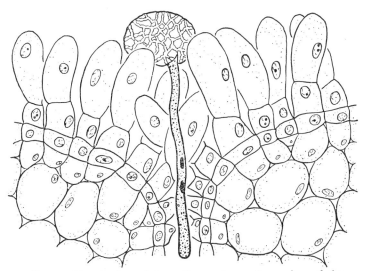

FIG. 17. Germination of the pollen grain on the surface of the stigma and the growth of the pollen tube down through the tissues of the style. The second pollen grain division has occurred and within the tube lie the two sperm nuclei.

visiting insects. The pollen may come into contact with a stigma, which is the receptive region of the female reproductive organ and here it is stimulated to germinate. Each germinating pollen grain produces a tiny tube which is capable of growing down through the tissues of the style (Fig. 17).

The pollen grain is merely a carrier of the generative nucleus from which are derived two **sperm nuclei** which function in fertilisation. These two nuclei arise from the **second pollen mitosis**. The generative nucleus sometimes divides in the pollen grain and the two sperm nuclei then move into the pollen tube. In some species, however, the generative nucleus itself moves into the pollen tube before undergoing this mitosis. Of the many pollen tubes which grow down the style usually only one passes into each ovule. The two sperm nuclei pass to the end of the tube which then bursts so that they are discharged into the embryo-sac. Fertilisation now takes place. Before a discussion of fertilisation it is essential to describe the development and structure of the embryo-sac.

The time that elapses between the beginning of meiosis and the formation of mature pollen grains differs considerably from one plant species to another. *Lilium regale*, for example, takes 18–20 days to progress from the beginning of meiotic prophase to the formation of tetrads and a further 15–20 days can elapse before pollen grains are shed from the ripe anthers.

PLATES XVII to XL

MEIOSIS AND THE DEVELOPMENT OF THE POLLEN IN *Lilium regale*

(Feulgen staining has been used to demonstrate the chromosomes. As a consequence, nucleoli, spindles and cytoplasmic bodies are not visible in these photographs.)

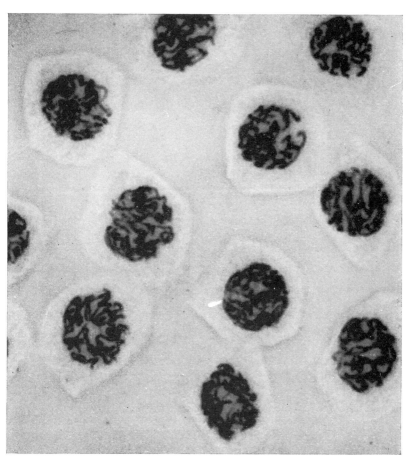

PLATE XVII

Nuclei synchronised at prophase of a pre-meiotic mitosis in a young anther (×550).

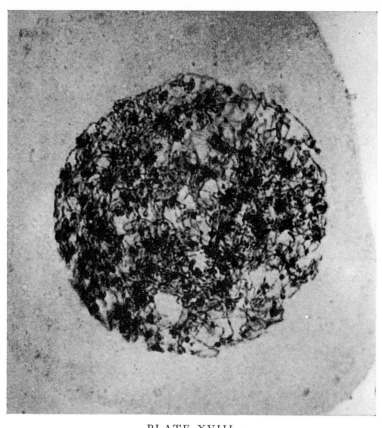

PLATE XVIII

Leptotene in a pollen mother cell. The nucleus is composed of 24 greatly extended chromosome threads (× 1700).

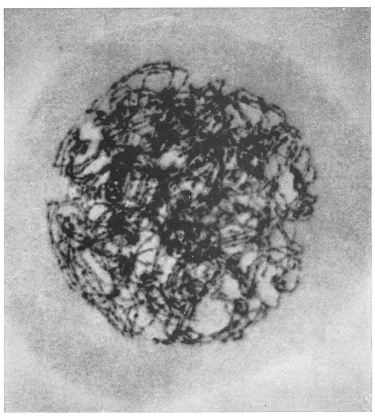

PLATE XIX

Zygotene. The chromosome threads have begun to associate in homologous pairs at various places along their length. Note the beaded appearance due to the presence of chromomeres (× 1700).

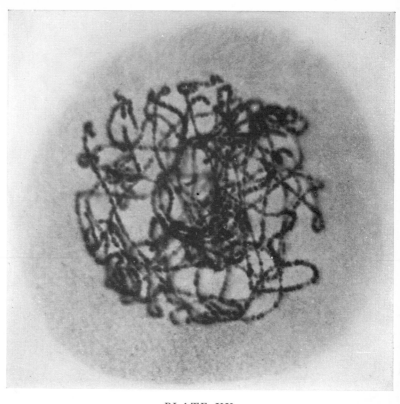

PLATE XX

Pachytene. Pairing of the homologous chromosomes is complete. The chromosomes have contracted and thickened and the chromomeres are now more obvious (× 1700).

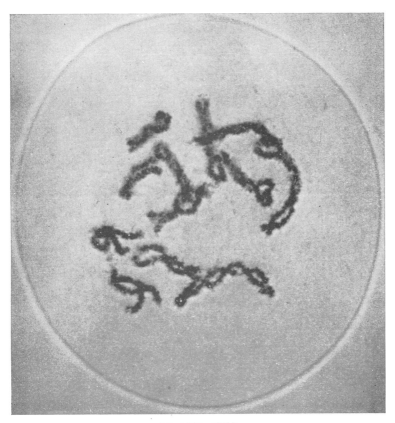

PLATE XXI

Diplotene. The chromosomes have contracted still further so that 12 bivalents are distinguishable. The attraction between the paired chromosomes of each bivalent has lapsed except where they are held together by one or more chiasmata. Note the loops between consecutive chiasmata (× 1700).

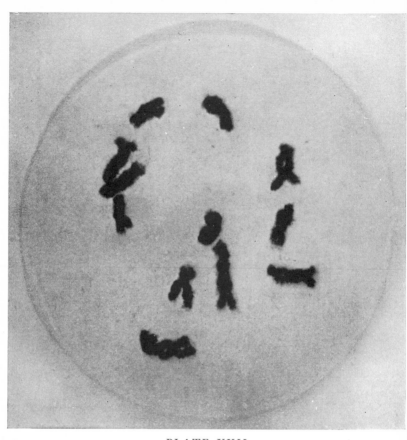

PLATE XXII

Diakinesis. The 12 bivalents have contracted further (× 1700).

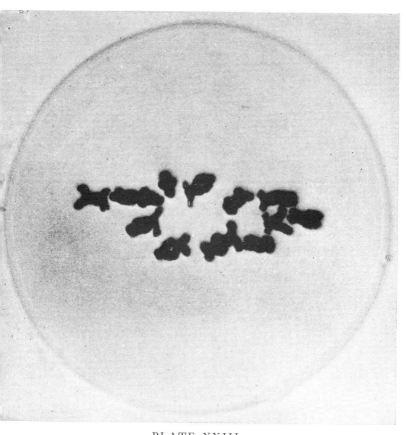

PLATE XXIII

First metaphase. The bivalents have become orientated upon the
equator of the spindle (× 1700).

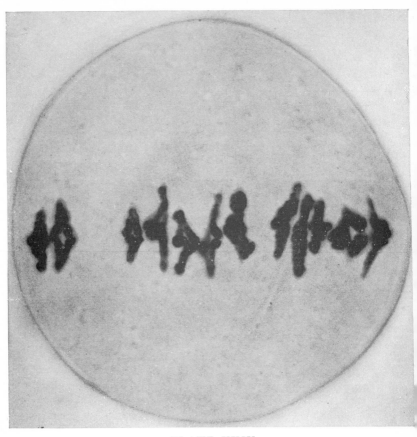

PLATE XXIV

First anaphase. Owing to centromere movement the half-bivalents are moving towards opposite poles of the spindle and the chiasmata are slipping apart (× 1700).

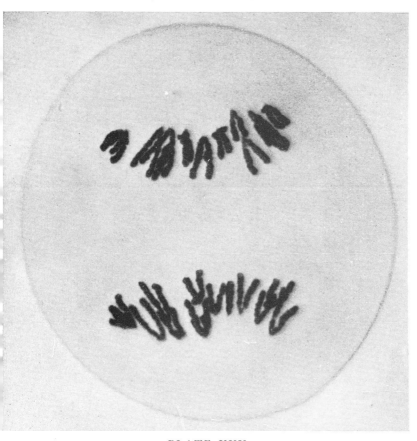

PLATE XXV

First anaphase. Separation is now complete and there are 12 chromosomes moving to each pole (× 1700).

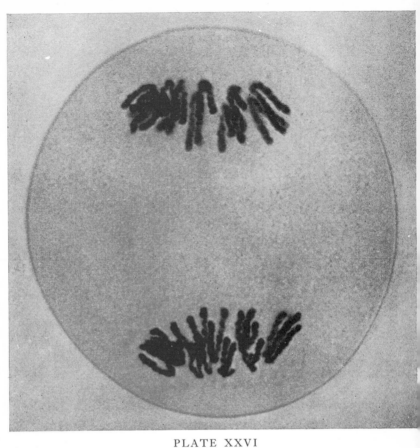

PLATE XXVI
First anaphase. The chromosomes have almost reached the poles
(× 1700).

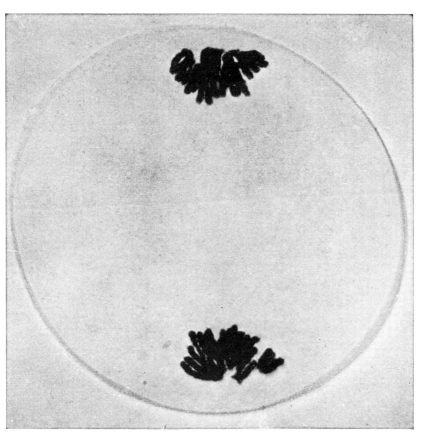

PLATE XXVII

First anaphase. The chromosomes having reached the poles now begin to form two compact groups (× 1700).

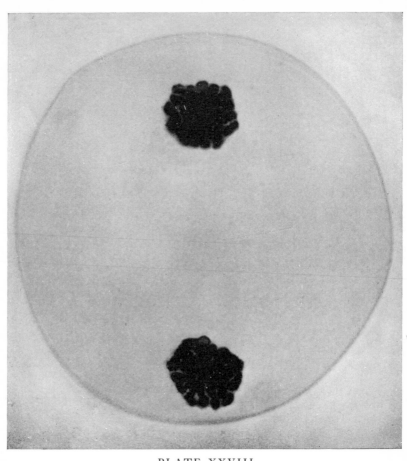

PLATE XXVIII

First telophase. The chromosomes have formed two nuclei (× 1700).

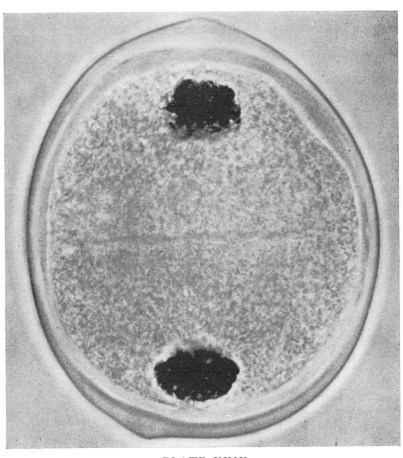

PLATE XXIX

First telophase. A new cell wall is beginning to form across the middle
of the cell to separate the two nuclei (× 1700).

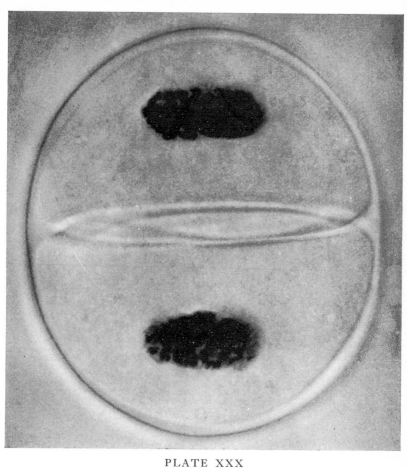

PLATE XXX

Interphase. Wall formation is complete and a dyad has been formed.
Each nucleus is composed of 12 chromosomes (× 1700).

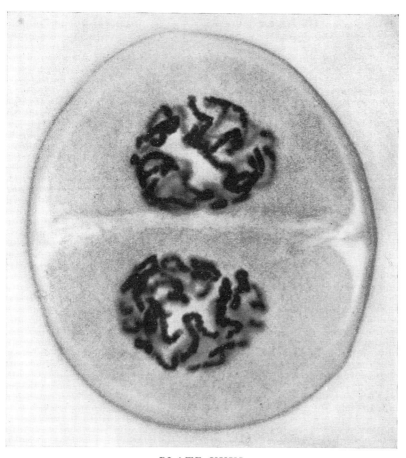

PLATE XXXI

Second prophase. The 12 chromosomes are now visible in each nucleus

(× 1700).

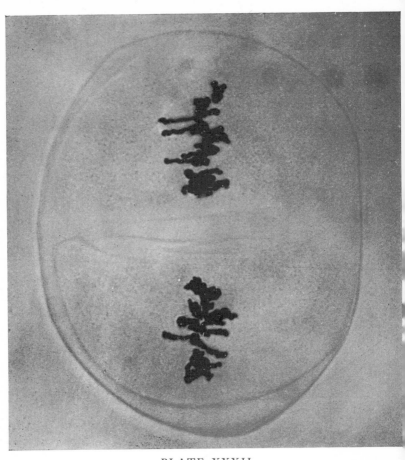

PLATE XXXII

Second metaphase. A spindle has formed in each cell of the dyad and the chromosomes have become arranged on the equators (× 1700).

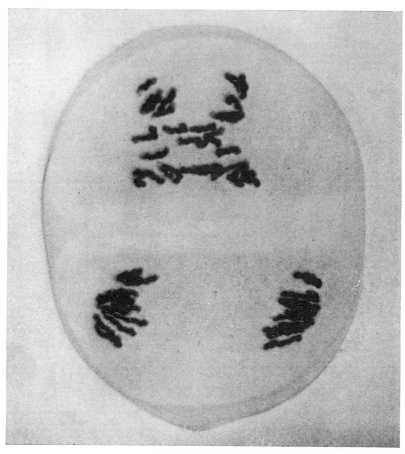

PLATE XXXIII

Second anaphase. The centromeres have divided and the chromatids are separating. The two cells of the dyad are not synchronised (× 1700).

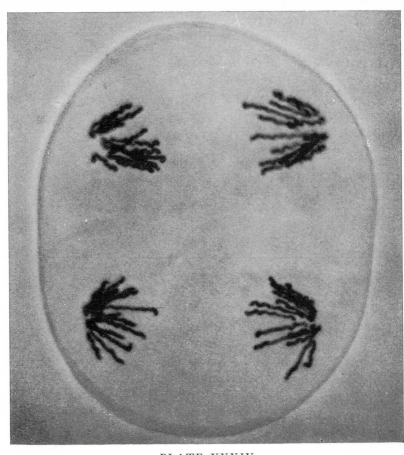

PLATE XXXIV

Second anaphase. An example of two synchronised cells of a dyad in which anaphase movement is nearly complete (× 1700).

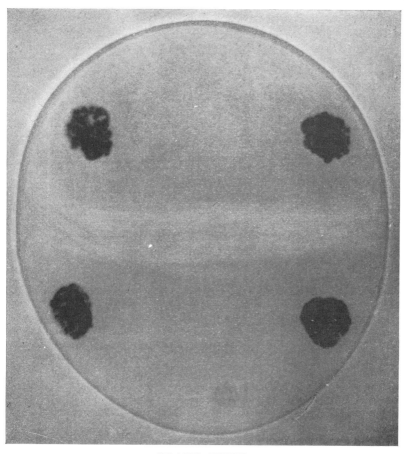

PLATE XXXV

Second telophase. The chromosomes have reached the poles and four
nuclei have been formed (× 1700).

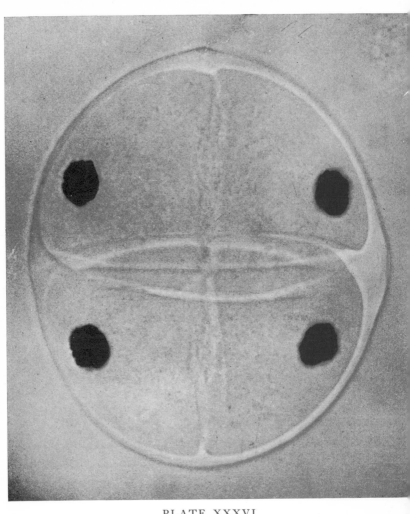

PLATE XXXVI
Second telophase. A new cell wall is developing (× 1700).

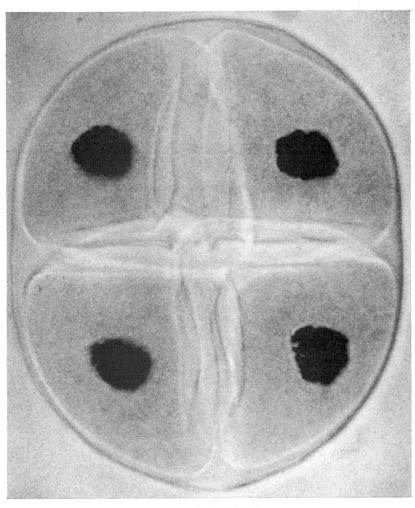

PLATE XXXVII

The tetrad. Four haploid cells have been formed. Each is a potential
pollen grain (× 1700).

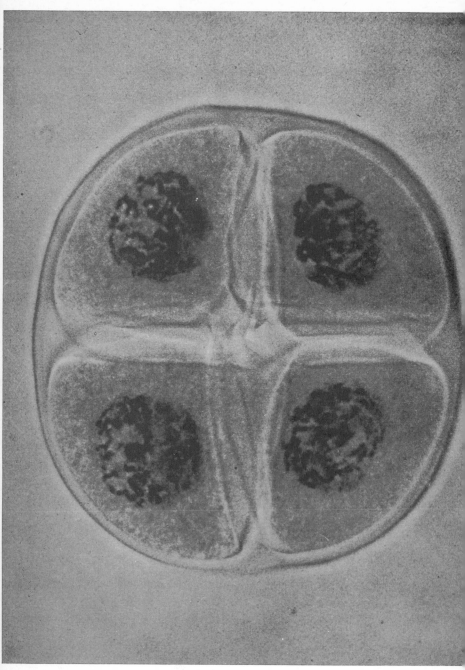

PLATE XXXVIII

The tetrad. The young pollen grains are developing (× 1700).

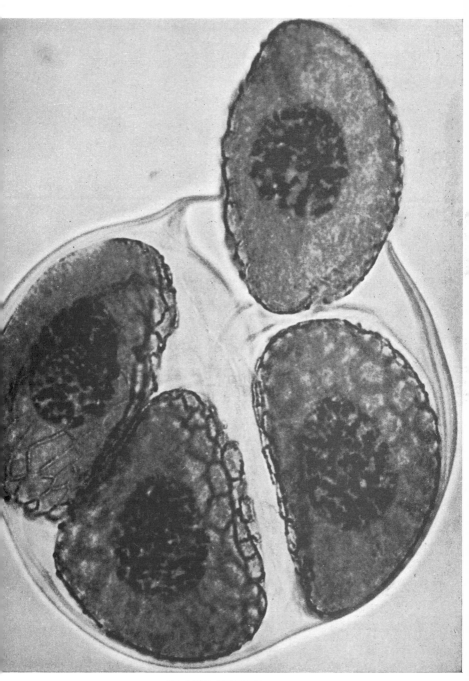

PLATE XXXIX

Young pollen grains. The four young pollen grains forming the tetrad are beginning to separate (× 1700).

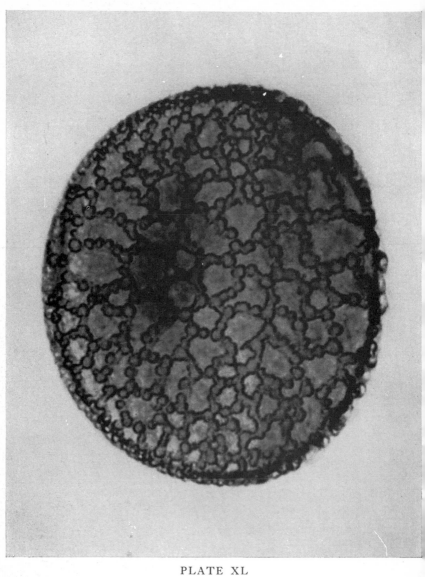

PLATE XL

Pollen grain. The wall has developed its characteristic patterning and
the pollen grain is nearing maturity (×850).

(ii) *The Formation of Embryo-Sacs*

Each female cell, or embryo-sac, develops from one **embryo-sac mother cell** embedded in the tissues of a young ovule. This cell enlarges and frequently elongates so that it finally becomes much larger than a pollen mother cell at a similar stage in development. The development of the female germ cell does not occur in step with that of the male cells but usually the timing is such that there is eventually synchronisation between the formation of the mature male and female gametes. Meiosis in the embryo-sac mother cell is similar to that in the pollen mother cell and also results in the formation of four genetically different haploid nuclei (Plates XLI to XLVI). This is the **primary four-nucleate stage** and it is comparable with the tetrad stage in the pollen mother cell. The way in which the embryo-sac continues its development varies considerably in different groups of plants. One, two, or even all four of the haploid nuclei may be concerned in the formation of a mature embryo-sac. Ultimately this can consist of from four to sixteen nuclei, the commonest number being eight. The final number of nuclei and their arrangement are determined by mitosis and by the migration of nuclei through the cytoplasm. In addition some of these nuclei fuse in a special way. For example, if two *haploid* nuclei undergo mitosis close together the chromosomes from both nuclei become orientated on one and the same spindle when the nuclear membranes finally break down. The daughter chromosomes, or chromatids, separate to form two nuclei each of which is then *diploid*. If, as happens in *Lilium*, more than two nuclei are involved in a fusion the embryo-sac will contain not only haploid nuclei but also nuclei with higher multiples of the chromosome number (Plates XLVII and XLVIII and Fig. 18). Only a few of these nuclei are functional in the process of fertilisation and the most important is the **egg nucleus** which, of course, is normally haploid in a diploid plant. There are also two or more **polar nuclei** which function in fertilisation and they may be haploid, diploid or even polyploid according to the type of embryo-sac development. The other nuclei, whose functions are unknown, are the **synergids** which often lie close to the egg nucleus, and the **antipodals** which lie at the opposite end of the embryo-sac. The type of embryo-sac developed in *Lilium regale* is illustrated in Fig. 18.

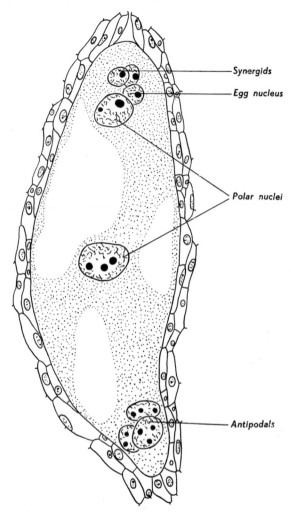

Synergids

Egg nucleus

Polar nuclei

Antipodals

FIG. 18. Diagram of a mature embryo-sac, of the type developed in *Lilium regale*, just before fertilisation. The haploid egg nucleus will fuse with a haploid sperm nucleus to form a diploid embryo. The two polar nuclei (one haploid and the other triploid) will together fuse with a different haploid sperm nucleus to form the pentaploid endosperm which divides abundantly and provides nutrition for the developing embryo.

PLATES XLI to XLVIII

MEIOSIS AND THE DEVELOPMENT OF THE EMBRYO-SAC IN
Lilium regale

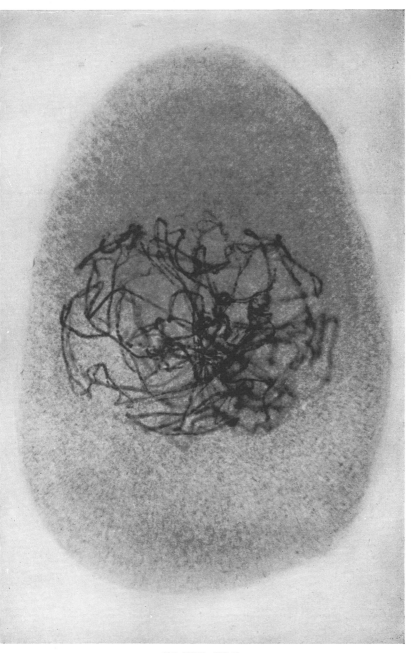

PLATE XLI
Zygotene in an embryo-sac mother cell (× 800).

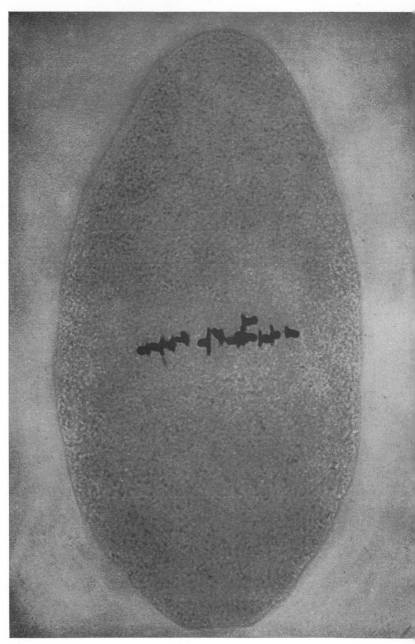

PLATE XLII
First metaphase (× 800).

82

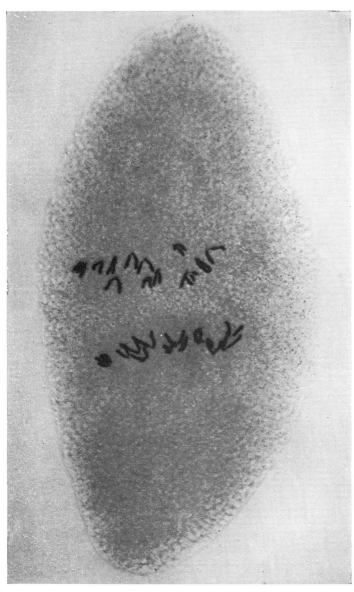

PLATE XLIII
First anaphase (× 800).

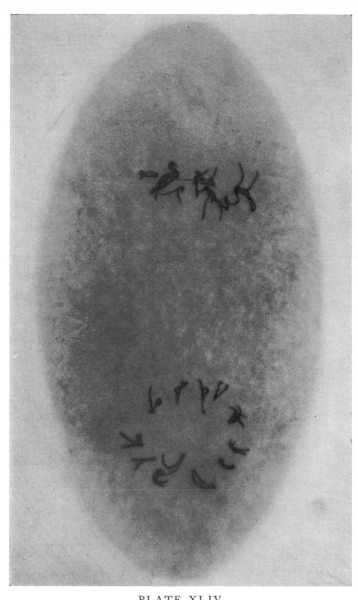

PLATE XLIV
Second metaphase (× 800).

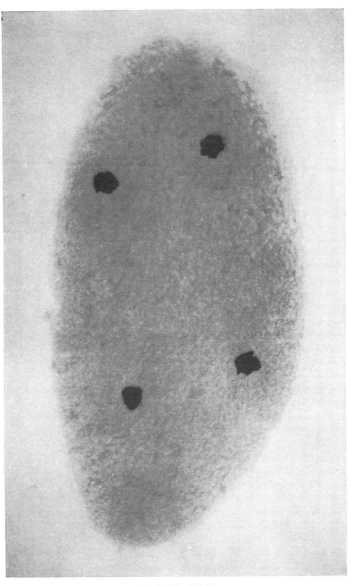

PLATE XLV
Second telophase (× 800).

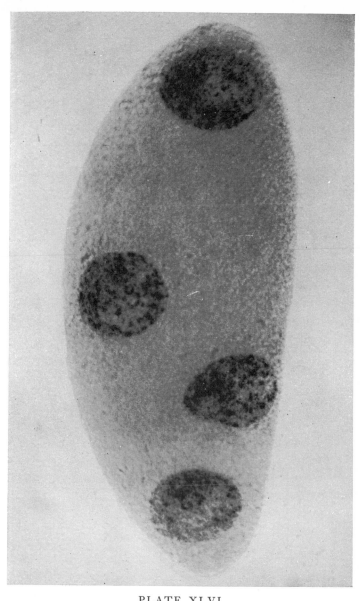

PLATE XLVI
Primary four-nucleate stage (× 800).

86

PLATE XLVII

Metaphases of mitosis in the four primary haploid nuclei with three sharing a common spindle. This will give rise to two haploid and two triploid nuclei (× 800).

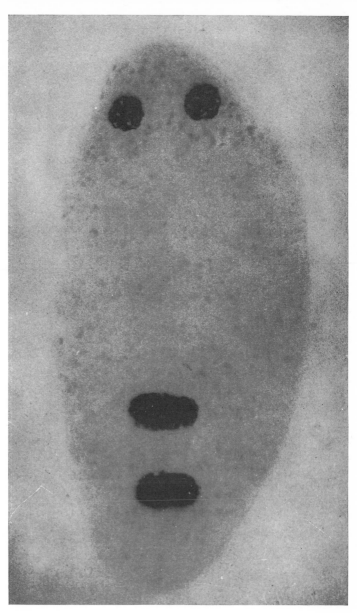

PLATE XLVIII

Secondary four-nucleate stage. One further mitosis of the two haploid
and two triploid nuclei will give rise to a mature embryo-sac with four
haploid and four triploid nuclei (see Fig. 16) (×800).

(iii) *Fertilisation and the Formation of Seed*

The pollen tube grows down the style and into the ovule where it ruptures allowing the two sperm nuclei to reach the embryo-sac. These sperm nuclei then pass through the cytoplasm and one comes into contact with the egg nucleus and the other with the polar nuclei.

There are conflicting reports concerning the behaviour of these closely associated male and female nuclei. Some observations suggest that one sperm nucleus first becomes intimately associated with the egg nucleus and eventually unites with it to form the diploid nucleus of the zygote. Others suggest that the two nuclei remain closely associated without fusing until both nuclei enter mitosis simultaneously. Then only one spindle is formed and the chromosomes from the male and female nucleus become arranged on it and restore the diploid number. If fusion takes place at mitosis in this way the formation of the zygote occurs, and at the same time its primary division, which results in the formation of the first two cells of the **embryo**.

Conflicting observations have also been made in connection with the union of the other sperm nucleus with the polar nuclei. The result of this union, however it may take place, is to form the **endosperm** which gradually increases in size by repeated divisions of its polyploid nuclei (Fig. 19). In *Lilium* the endosperm is pentaploid through the fusion of a haploid sperm nucleus, a haploid polar nucleus and a triploid polar nucleus. The function of the endosperm is to nourish the embryo in its early stages of development which is by no means haphazard because early differentiation of the cells ensures that the embryo shall grow in a regular way. The growing embryo with its nutritive endosperm and the surrounding tissues of the ovule form the young developing seed. In some plants the endosperm is absorbed by the embryo but in others, such as *Lilium*, the endosperm becomes multicellular and forms a large part of the mature seed when it is finally shed from the plant.

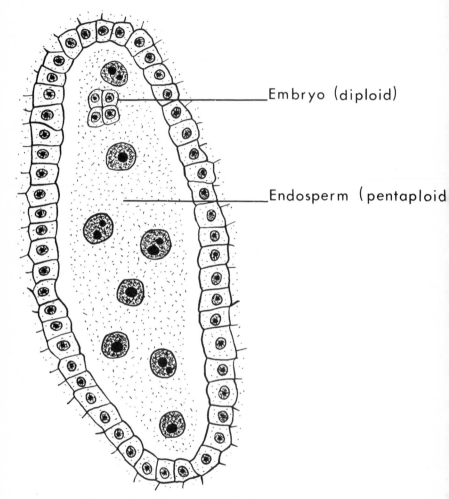

Embryo (diploid)

Endosperm (pentaploid

FIG. 19. An early stage in the development of the seed showing the diploid nuclei of the young embryo and the pentaploid nuclei of the nutritive endosperm in *Lilium regale*.

4

MITOSIS, MEIOSIS AND HEREDITY

So far, the part played by mitosis in the development of a mature plant and the equally important part played by meiosis in the formation of its reproductive cells have been described (Fig. 20). The emphasis throughout has been on

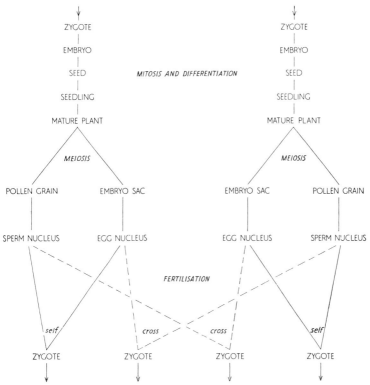

FIG. 20. Diagram illustrating some of the stages in the development and reproduction of two flowering plants and how new generations may arise through either self- or cross-fertilisation.

descriptions of chromosome behaviour and how this behaviour is related to the maintenance of the genetic constitution. Now, however, brief reference must be made to some of the deviations which are known to occur normally within an individual and consideration given to some of the genetic consequences of any deviations that may be inherited.

At the time of fertilisation, when the chromosomes from the sperm and the egg nucleus together form the nucleus of the zygote, the genetic constitution of that zygote, and of the individual into which it will develop, is irrevocably established. The precision of regular mitotic divisions ensures a basic and essential genetic continuity. There are, however, some cells distinct from the germ line which do differ. For example, it is normal in plant development for the nuclei of some cells to contain multiples of the original number of chromosomes. These polyploid cells can be of two kinds: one may arise from successive periods of DNA synthesis without cell division; the other, for example, may arise from the failure of chromatids to move apart at anaphase so that the resulting single nucleus comes to contain double the number of chromosomes.

When a plant is propagated vegetatively, that is by means of bulbs, corms, rhizomes, stolons and so on, the chromosomes and genes are passed on entire and unchanged. Thus parents and offspring are identical both in appearance and in their response to an environment. A population of plants so formed is known as a **clone**. Although vegetatively reproducing plants may be at an advantage in not having to rely upon favourable conditions for the production and germination of seed, they have a very low degree of genetic variability compared with sexually reproducing plants. This renders them static from an evolutionary point of view and the extent of their variability depends on changes in chromosome number and on **mutations** which are changes in the structure or composition of the genes. Mutations and changes in chromosome number do occur spontaneously with very low frequency but can also be induced by certain chemicals and, particularly in the case of mutation, by radiations. Changes in chromosome number are not restricted to polyploidy because, very occasionally, **aneuploid** plants are found which have chromosomes extra to, or less than, the diploid complement.

Sexually reproducing plants are known to be capable of producing offspring which differ genetically from one another and from their parents. It is during meiosis that genes become recombined by crossing-over and finally distributed in new and varying combinations to the germ cells. As in vegetatively reproducing plants polyploidy, aneuploidy and mutation also occur. Together with the recombination of genes resulting from crossing-over in meiosis, these give wide scope for variation.

New plants are continually arising, therefore, which may have small genetic differences although they still retain the characteristic appearance of the species or taxonomic group to which they belong. Thus there is a mechanism whereby the variation of the plant population through consecutive generations provides for the adaptation of that population, through selection, to new and changing environments.

APPENDIX

Although cytological and photographic techniques are described in one of the books listed in the References, brief details of the techniques recommended for looking at chromosomes in *Lilium regale* are given below.

Root-tip Meristems

1. Treat excised roots in 0.05% colchicine for 5–6 h at 20–25 °C.
2. Fix in La Cour's 2BD for 1 h.
3. Add an equal volume of 1% aqueous chromium trioxide to the fixative and allow to stand for 2 h.
4. Wash thoroughly in distilled water for 20 min.
5. Treat for 5–10 min in freshly made oxalate-peroxide solution (equal volumes of a saturated solution of ammonium oxalate in distilled water and 20 vol. hydrogen peroxide).
6. Rinse in distilled water.
7. Hydrolyse in NHCl at 60 °C for 12 min.
8. Stain in Feulgen's reagent for 2 h.
9. Macerate meristem in 45% acetic acid on slide, apply coverslip and squash.
10. Slides can be examined immediately and then, if required, made permanent. (Stage 1 should only be included if a detailed study of metaphase chromosomes is required.)

Pollen Mother Cells

1. Fix whole anthers for 1–2 h in acetic-alcohol (one volume of glacial acetic acid with three volumes of absolute alcohol).
2. Pass anthers through an alcohol series to water: 80%, 60%, 40%, 20% alcohol and water, 10 min in each.
3. Hydrolyse in NHCl at 60 °C for 6 min.
4. Stain in Feulgen's reagent for 2 h.
5. Dissect out anther contents in 45% acetic acid on a slide, apply cover-slide and squash.
6. Slides can be examined immediately and then, if required, made permanent.

Embryo-sac Mother Cells

1. Dissect ovules from ovary and fix over-night in Carnoy's fixative (one volume of glacial acetic acid plus three volumes of chloroform plus six volumes of absolute alcohol).
2. Pass ovules through an alcohol series to water (see Stage 2 in the schedule for pollen mother cells).
3. Hydrolyse in NHCl at 60 °C for 8 min.
4. Stain in Feulgen's reagent for 2 h.
5. Place four of the stained ovules in a large drop of 45% acetic acid on one slide and apply a cover-slip.
6. Observe ovules under low magnification and tap cover-slip repeatedly with point of needle until embryo-sac mother cells are released. Use filter paper to remove excess acetic acid.
7. Slides can be examined immediately and then, if required, made permanent.

Photographs

A green filter was used in the microscope lamp to increase contrast and the photographs were taken with a simple eyepiece camera. Half-tone panchromatic plates were used and developed by standard procedures.

REFERENCES

The following books are recommended for reference and further reading:

ABERCROMBIE, M., HICKMAN, C. J. and JOHNSON, M. L., *A Dictionary of Biology*. Revised ed., 1969. (Penguin, London).

AUERBACH, C., *The Science of Genetics*, Revised ed., 1970. (Harper and Row, New York).

BONNER, D. M., *Heredity*, 2nd ed., 1964. (Prentice-Hall, New Jersey).

BURNHAM, C. R., *Discussions in Cytogenetics*, 1962. (Burgess, Minneapolis).

CLOWES, F. A. L. and JUNIPER, B. E., *Plant Cells*, 1968. (Blackwell, Oxford).

DARLINGTON, C. D. and LA COUR, L. F., *The Handling of Chromosomes*, 5th ed., 1969. (George Allen and Unwin, London).

ESAU, K., *Plant Anatomy*, 2nd ed., 1967. (Wiley).

JOHN, B. and LEWIS, K. R., *The Meiotic System*, 1965, Protoplasmatologia Band VI, F1. (Springer-Verlag, Vienna).

KENDREW, J., *The Thread of Life*, 1966. (Harvard University Press, Cambridge, Mass.).

SCIENTIFIC AMERICAN, *The Living Cell*, Introduced by KENNEDY, D., 1965. (W. H. Freeman, San Francisco).

LEWIS, K. R. and JOHN, B., *Chromosome Marker*, 1963. (J. and A. Churchill, London).

LEWIS, K. R. and JOHN, B., *The Matter of Mendelian Heredity*, 1964. (J. and A. Churchill, London).

McELROY, W. D. and SWANSON, C. P., *Modern Cell Biology*, 1968. (Prentice-Hall, New Jersey).

PAUL, J., *Cell Biology*, 2nd ed., 1967. (Heinemann, London).

RIEGER, R., MICHAELIS, A. and GREEN, M. M., *A Glossary of Genetics and Cytogenetics*, 1968. (Springer-Verlag, Berlin).

SRB, A. M., OWEN, R. D. and EDGAR, R. S., *General Genetics*, 1965. (W. H. Freeman, San Francisco).

SWANSON, C. P., *The Cell*, 3rd ed., 1969. (Prentice-Hall, New Jersey).

WATSON, J. A., *Molecular Biology of the Gene*, 1965. (W. A. Benjamin, New York).

WHITE, M. J. D., *The Chromosomes*, 5th ed., 1963. (Methuen, London).

WHITEHOUSE, H. L. K., *Towards an Understanding of the Mechanism of Heredity*, 2nd ed., 1969. (Edward Arnold, London).